J. Langton H., R. Savill H.

Metrical Translations and Other Verses

J. Langton H., R. Savill H.

Metrical Translations and Other Verses

ISBN/EAN: 9783337312572

Printed in Europe, USA, Canada, Australia, Japan

Cover: Foto ©berggeist007 / pixelio.de

More available books at **www.hansebooks.com**

METRICAL TRANSLATIONS

AND OTHER VERSES.

METRICAL TRANSLATIONS
AND OTHER VERSES

BY

J. LANGTON H. AND R. SAVILL H.

LONDON:
PRINTED AT THE CHISWICK PRESS.
1884.

CONTENTS.

	PAGE
WINTER Piece	3
On Milton's "Paradise Lost"	5
To an Amaranthus Blowing late in Autumn	8
Love and Folly	11
From "Mathematical Questions"	13
The Animal in the Moon	15
The Grasshopper and the Ant	18
Description of Calypso's Island	23
Calypso's "Good-Night" to Telemachus	25
Apollo and the Thessalian Shepherds	27
The Marine Festival	29
The Delusion of Athamas	31
Antiope	35
The Approach of Old Age	37
The Barbarian Ambassadors to Idomeneus	39
Cleanthes and Pholoë	40
The Dead Body of Hippias	42
Mentor assuming the Attributes of Minerva	43
The Sculptor and the Statue of Jupiter	45
Distrust	47
Cain	50
The Translation of Enoch	52
The Death of Moses	53
The Song of Deborah and Barak	55

CONTENTS.

	PAGE
The Denouncement of Judgment upon David	58
The XCVII. Psalm	61
Solomon's Prayer	63
A Reverie in Cumberland	65
A Young Lady's Farewell to Peterborough	68

METRICAL TRANSLATIONS, ETC.

By J. LANGTON H.

A WINTER PIECE.

WITHOUT control stern Winter rages now,
    And subject Nature to his sway submits;
Tempests and drizzling rains, and frost and snow,
    Attendant in his train, he deals by fits.

How desolate the scene! the fields are bare;
    The trees no more their leafy dress display;
Clouds veil the sun, and piercing is the air,
    And hush'd is every songster's tuneful lay.

How chang'd is all since Summer lately threw
    His verdant robe the smiling landscape o'er!
Now all seems cold and barren to the view,
    And woods, hills, valleys, please the eye no more.

Impatient, oft we at the change repine;
    But let us not despairingly complain;
For Winter soon his empire shall resign,
    And joyous Spring and Summer laugh again.

Meanwhile this useful lesson we may learn;
   As Winter meliorates the exhausted soil,
By crumbling frost and plenteous rains, in turn,
   That Autumn's bounty crown the labourer's toil,

So is affliction, Winter of the soul,
   In mercy by the all-wise Creator given,
To melt our hearts, our vain desires control,
   And fit us for the eternal Spring of heav'n.

ON MILTON'S "PARADISE LOST."

FROM THE LATIN OF SAMUEL BARROW, M.D.

WHOE'ER thou art that read'st this heavenly song,
   By the great Milton writ, I ask of thee,
Doth not to every part the praise belong
   Of perfect judgment, true sublimity?
To what a height his Muse adventurous soar'd!
   What lofty themes composed his daring strains!
The origin of all things he explor'd;
   The fate awaiting all, this book contains.
Creation's wonders, too, he did declare;
   The mysteries of this terrestrial frame;
The earth, the boundless sea, the ambient air,
   Dark Erebus, and hell that vomits flame;
Whate'er on earth or in the ocean dwell,
   Or the bright regions of celestial skies,
Or in the drear abodes of darkest hell,
   And what is shut as yet from mortal eyes;
Chaos unbounded, a God infinite,
   And great expression or compare above;

If that is great that will no bound admit—
　In Christ towards men conciliating love.
Who would have thought or hoped themes so sublime
　In poetry could ever thus succeed?
Yet favour'd Britons, in our happy time,
　May in their bard such heavenly raptures read.
War, too, he sung, gender'd by Satan's pride;
　Angelic hosts fill'd heaven with dire alarms.
What mighty generals on either side!
　Michael and Satan stand opposed in arms.
How terrible the weapons that they wield!
　How dreadful is each adverse trumpet's clang!
Angels the warriors; heav'n the battle-field:
　The skies with uproar and confusion rang.
How Satan rages! how aloft he rears
　His dreadful weapons; as he strides the field,
To Michael scarce inferior he appears;
　Leads on his squadrons, and disdains to yield.
The mountains from their bases torn, are thrown
　Like darts by either host, and fire they rain.
Olympus stands in doubt which it shall own
　Victorious on the long contested plain.
But now in air Messiah's ensigns blazed;
　Upon a living chariot he rode;
Sublime he moves, the thunder high upraised
　In his right hand, arms worthy of the God!
The roaring wheels cast lightning as they roll,

And withering fire the blazing chariot darts;
Hoarse thunders growl, and sound from pole to pole,
   Dismay and anguish seize the rebels' hearts.
They fly confounded: fiery darts pursue,
   And drive them to the verge of opening hell:
Themselves to 'scape the wrath, they downward threw;
   For ever there in darkness shall they dwell.
Yield Greece and Rome; the bays let Milton wear;
   Ye modern poets, yield; for if his theme,
With Homer's or with Virgil's we compare,
   But frogs or grasshoppers their subjects seem.

TO AN AMARANTHUS BLOWING LATE IN AUTUMN.

FROM THE FRENCH OF M. CONSTANT DUBOIS.

OUR gardens no longer are gay
    With the flowers which they lately display'd:
I have witness'd them droop and decay,
    Though soft zephyrs around them have play'd.
Flora's beauties, alas! are no more:
    How short is the time that they last!
Thus quickly our pleasures are o'er,
    So soon all enjoyments are past.

Thee, sweet Amaranthus, I see;
    To lessen my grief dost thou come:
Thy fragrance exhaling for me,
    And unfolding thy delicate bloom.
Thus when storms of adversity blow,
    And o'er our bereavements we grieve,
Kind friendship will solace our woe,
    And in all our distresses relieve.

TO AN AMARANTHUS.

So beauteous thy lovely attire,
   So charming to me dost thou seem,
Thou inspirest my languishing lyre;
   I cannot but make thee my theme.
Last tribute that Flora bestows,
   Which the Goddess lets fall from her zone
That we may not forget when she goes,
   What delights with her presence have flown.

So when a loved friend doth depart
   On a journey to some distant land,
His still gazing eyes speak his heart,
   As he waves us farewell with his hand.
Tender language, though mute, which declares,
   As plainly as words could convey,
" The last pledge of the love my heart bears;
   Which it still shall retain whilst away."

When a few more short years shall have flown,
   And my summer of life has an end,
Ah, then my old age thou shalt crown,
   And thou shalt to me be a friend.
Thine image shall bring to my mind
   Sweet remembrance of happier hours;
My old age to my youth shall seem join'd,
   When I greet thee, thou last of the flowers.

Thy smile, lovely flow'ret, shall cheer,
    When I gaze on thy beautiful form;
And shall give me new courage to bear
    Each adverse and pitiless storm.
If an aspect of gladness Spring bears,
    The Autumn of life, too, is sweet :
I welcome ye, my silver hairs :
    Anacreon his joyfully did greet.

Does happiness solely depend
    On either youth, manhood, or age ?
Youthful schemes rarely meet their wish'd end :
    Disappointment attends on each stage.
How oft does age follow the bier
    On which youth to the tomb is convey'd !
How oft shed the warm, gushing tear,
    O'er the grave where an infant is laid !

LOVE AND FOLLY.

FROM THE FRENCH OF LA FONTAINE.

LOVE is a mystery, no doubt,
   And I, for one, can't make it out.
   His darts, his flames, his infancy,
Were never understood by me.
My sole intent is to explain,
In the best manner that I can,
Why this omnipotent God is blind,
And show herein the Fates were kind,
And favoured man in this decree:
What power had Love if he could see!
Lovers on this may judgment pass;
I 'll not decide so nice a case.

  It chanced upon a certain day
That Love with Folly was at play.
Love was not then deprived of eyes;
Between them angry words arise,
And Love proposes that the Gods
Be summon'd from their high abodes

To try the matter in debate.
Impatient Folly would not wait,
But struck at Love with rage and spite,
And quite deprived the God of sight.
Venus for vengeance rends the skies
With shrieks and loud maternal cries.
Jupiter, Nemesis, and all
Who judge in hell, come at her call.
To them she represents the case;
How infamous the act, how base!
Her son in a defenceless state;
No punishment could be too great!
And they in justice must requite
Her darling boy for loss of sight.
They duly weigh'd the grave affair,
Considering every point with care;
And thus at last the Court decided:
" Love, evermore, shall be by Folly guided."

FROM "MATHEMATICAL QUESTIONS."

 LAWYER, Physician, and Priest
    Once happen'd together to dine ;
And having concluded the feast,
    Were chatting and sipping their wine.

After drinking full many a toast,
    (How many, the muse did not count)
Each began to the others to boast
    Of his riches the wondrous amount.

They found that the sum of their gains
    Proportion harmonical bore ;
Of course this kept out of their brains
    Thoughts of envy at each other's store ;

That the total amount of their wealth
    Was one hundred four thousand pounds ;
Of which the restorer of health,
    Who in riches the most did abound,

Had half what the lawyer possess'd,
   Plus four times the sum that remain'd,
By deducting that own'd by the priest,
   From the sum that the lawyer had gain'd.

I beg from the data I give,
   You will find what to each did belong;
And with hope you'll successfully strive,
   I here make an end of my song.

It is well known that the reciprocal of numbers in harmonical progression, are in arithmetical progression. Let x = the physician, and v the common difference of their reciprocals; then x, $\frac{1}{\frac{1}{x}+v}$, $\frac{1}{\frac{1}{x}+v'}$ will represent the gains of the physician, lawyer, and priest respectively.

$\therefore x + \frac{1}{\frac{1}{x}+v} + \frac{1}{\frac{1}{x}+v} = a$ ...... (1), and (per question) $\frac{1}{2\frac{1}{x}+v} + \frac{4}{\frac{1}{x}+v} - \frac{4}{\frac{1}{x}+2v} = x$, or $\frac{4\frac{1}{2}}{\frac{1}{x}+v} - \frac{4}{\frac{1}{x}+2v} = x$. Divide both sides of the last equation by x, and the result is $\frac{4\frac{1}{2}}{1+vx} - \frac{4}{1+2vx} = 1$. Hence $v^2 x^2 - vx = -\frac{1}{4}$, and consequently $x = \frac{1}{2v}$. By substituting this value of x in equation (1), we have $\frac{1}{2v} + \frac{1}{3v} + \frac{1}{4v} = a$. Hence $v = \frac{13}{12a}$. $\therefore x = 48000$, $\frac{1}{3v} = 32000$, $\frac{1}{4v} = 24000$.

THE ANIMAL IN THE MOON.

From the French of La Fontaine.

SOME of the philosophic tribe have said,
Man by his senses always is misled;
Others maintain they are a certain guide,
And never lead us from the truth aside.
Each creed in part is true: without the aid
Of reason, blunders will be often made.
Those who their senses only will believe,
Appearances continually deceive.
The sun, that dazzling orb, that fount of light,
How small a space it occupies to sight!
But reason tells me it is nature's eye,
The ruler of the wondrous worlds on high.
I know 'tis vastly distant, and from thence
Infer its size, nor take my guide from sense.
I see its surface as a plane appear,
Though reason teaches me it is a sphere;
It seems to run the circuit of the sky,
And earth appears immoveable to lie;
But both appearances are false; the race
Is run by earth; the sun preserves his place.

I shut my eyes; a dismal blank ensues,
And lost are nature's bright and varied hues.
Yet though no longer to my sight reveal'd,
I know they still exist, but are conceal'd.
A stick however straight will bent appear
When introduced in water still and clear.
Thus, through disguise of sense, the active mind
Strives ever truth in its own form to find.

By the unlearn'd 'tis confidently said
On the moon's surface is a female head;
And various other wonders they declare;—
A man, an ox, an elephant are there.
They, in her inequalities of face,
All kinds of strange, misshapen monsters trace;
Nor to the uninstructed of mankind
Such silly fancies are alone confined.
By way of illustration I will tell
What once a sage astronomer befell:
His telescope he pointed towards the moon,
And something wondrous was apparent soon.
An animal of most prodigious size,
Within its orb met his astonished eyes;
And all who used the tube observed the same:
The marvel far and wide was spread by fame.
Some dread event 'twas thought to indicate;
Dissensions and convulsions in the State;

THE ANIMAL IN THE MOON.

Wars, pestilence, or famine.   Every ear
Was open to receive some tale of fear.
The monarch would himself the wonder view,
And find if this appalling tale were true.
But when the huge, terrific beast he saw
His breast was fill'd with wonder and with awe.
At length by accident was brought to light
The harmless cause of all the needless fright.
A mouse into the tube had found its way,
And thus occasion'd all this wild dismay.
But now the cause of wonderment was clear,
And mirth and laughter took the place of fear.

THE GRASSHOPPER AND THE ANT.

FROM THE FRENCH OF LA FONTAINE.

 GRASSHOPPER had spent the summer long
In idle sport, and never-ceasing song;
But when rude Boreas from the northern sky
Brought wintry storms, she found nor grub nor fly,
So raised to neighbour ant a piteous cry,
And humbly begg'd that from her garner'd store
Her wants might be relieved; nor ask'd for more
Than what might just sustain her life and strength
Till plenteous summer should arrive at length.
"When August comes, with interest I'll repay;
And take an insect's word for what I say."
Thus pray'd Miss Grasshopper. The ant, we know
Is not inclined to lend, much less bestow.
Here lies her fault. She to the suppliant's prayer
This answer gave: "What did you make your care

While summer suns were beaming?" "Night and day,"
Said Grasshopper, "I sang 'dull care' away."
"You sang; how happy and content you were!
Now dance, and once again get rid of care."

METRICAL TRANSLATIONS FROM FENELON'S TELEMACHUS, AND OTHER VERSES.

By R. Savill H.

DESCRIPTION OF CALYPSO'S ISLAND.

Book I.

SOFT zephyrs, playing round the grot, maintain'd
Delicious coolness, spite of Sol, who reign'd
In burning splendour: fountains murmur'd round,
And as along those verdant meads they wound,
Adorn'd with amaranths and violets, there
They fell in baths, as crystal pure and clear.
The neighbouring lawns like a green carpet lay,
Deck'd with a thousand flowers. In rich array
Appear'd a wood of those wide branching trees
Which bear the golden apple, and the breeze
Came loaded with the sweetest of perfume,
Borne from their blossoms. In perpetual bloom
They stood and crown'd the meads. The beams of day
Were from their shade excluded: there the lay
Of birds alone was heard, save where the brook,

Impetuous rushing from the rocks, forsook,
With foamy waves, the shade, and took its way,
With gentler course along the meads to stray.

   The grotto from on high o'erlook'd the deep,
Now as a mirror bright in tranquil sleep:
Anon, its billows beat the rocky shore
With idle rage, and wat'ry mountains roar.
But, turning thence, a river met the view,
Adorn'd with isles, where flow'ring lindens grew,
And poplars rear'd aloft their noble heads,
Which strove to pierce the clouds. Along the meads
The streams which form'd the islands seem'd to play;
Some with swift course pursued their crystal way;
Some gently flow'd, and others turn'd again,
As loath to leave this sweet, enchanting plain.

   Distant afar, blue mountains heave on high
Their mighty peaks, which soar into the sky,
And with fantastic forms delight the eye.
Here vine-clad slopes diversify the scene;
Thick clusters blush behind their leafy screen:
There like a garden spreads the smiling plain,
And trees of every fruit enrich the wide campaign.

CALYPSO'S "GOOD NIGHT" TO TELEMACHUS.

Book IV.

ALYPSO, motionless, with rapture fill'd,
The youthful hero heard. At length the hour
Of rest drew nigh, and thus the Goddess spake:
Let slumber now thy lengthen'd toils succeed.
Fear from thy breast be banish'd. All for thee
Here smiles propitious—Peace and every joy
Which Gods confer on men await thee here.
With rosy fingers when the gates of morn
Aurora in the golden east unbars,
And Sol's bright chariot, bursting from the wave,
Drives from his course the lesser lights of heav'n,
Then, my Telemachus, thou shalt resume
The history of thy wanderings.—Thou thy sire
In wisdom and in virtue hast excell'd.
Achilles, vanquisher of Hector; he
Who rose triumphant from the gulf of hell,
Great Theseus; e'en the mighty son of Jove,

Who rid the earth of monsters—all than thou
Less fortitude, less valour have display'd.
May night to thee, in sweet refreshing sleep
Consumed, seem swiftly to have pass'd away.
But I, alas! shall deem its flight too slow.
How shall I long to see thee, and to hear
Thy voice again! To make thee tell once more
What thou hast told, and what I know not yet
Beseech thee to unfold. With Mentor, then,
Thy heav'n-sent guide, to yon high grot repair:
I will invoke for thee the God of Dreams,
That he may shed his softest influence
O'er thy sleep-laden lids: bright visions send,
Which hovering round thee, may enchant thy sense
With images of bliss; and far aloof
Keep all which might untimely break thy rest.

APOLLO AND THE THESSALIAN SHEPHERDS.

Book II.

SHORN of his beams, an exile from the sky,
Apollo dwelt on earth, and fed the flocks
Of King Admetus.   Oft his warbling flute
Awoke the echoes of the rock and grove,
And swains, assembling in the breezy shade
Of waving lindens, by the fountain's flow,
With rapture heard the dulcet strains.
                                       He sung
The flowers which herald Spring; the balmy gales
Which play around her, and the youthful verdure
Budding beneath her footsteps.   Next he sung
The nights of Summer when descending dews
Refresh the thirsty earth and drooping flowers.
He sung the fruits of Autumn which reward
The labourer's toil, and Winter's gloomy reign,
When sprightly youths around the ruddy glow
Of festive torches, dance away the hours.

He sung the gloomy forests which adorn
The mountain's rugged steep ; the fertile vales,
Where rivulets, meandering, seem to sport
Amid the laughing meads.
                              He taught the swains
The charms of rural life, and they, whose nature
Was heretofore uncivilized and rude,
Became more happy, with their rustic pipes,
Than monarchs with their sceptres, for their lives
Were one continuous holiday. The groves
Rung with the song of birds, the dash of fountains,
The music of the gently waving boughs,
And shepherds' voices, raised in hymns of praise.

THE MARINE FESTIVAL.

Book VIII.

THE rowers were with bays and roses crown'd;
The stroke of oars kept measure to the sound
Of flutes: Achitoas awakes the lyre,
And with his voice celestial strains conspire.
That voice might bless the banquets of the skies,
And charm Apollo.   Round the vessel rise
Tritons and nereids, hoary Neptune's train,
And vast sea-monsters skim the surface of the main.
They quit their grottoes, humid and profound,
All fix'd in charm'd attention at the sound.
Phœnician damsels dance in fair array,
In snow-white vestures clad, and lovely as the day;
And ever and anon the sprightly sound
Of trumpets floated on the waves around.
The calmness of the sea, the silent night,
And the pale moonbeams, which, with trembling light,
Play on the waves; the deep celestial blue,
Thick sown with stars, presented to the view
A matchless scene.   Telemachus beheld;
With rapture at the sight his bosom swell'd.

Then Mentor swept the lyre : Achitoas' breast
Was filled with shame and envy, while the rest
Enraptur'd heard the soul-entrancing strain.
He sung of Jove, the sire of gods and men,
Who shakes creation with his awful nod:
Of Pallas, who, proceeding from the God,
To teach mankind forsakes the blest abode.
And so sublime the strain, that awe possest,
And sacred horror, every hearer's breast,
As if he stood on vast Olympus' height—
The Thunderer confest before his sight,
Whose awful eyes more dreadful light diffuse
Than e'en his burning bolts ; and then the muse
He bids unfold how young Narcissus died,
Cut off, untimely, in his youthful pride ;
Enamour'd of his own reflected charms,
He stands, and longs to catch the shadow in his arms ;
Still o'er the fount he bends, and pines with grief,
Till pitying gods afford him kind relief,
And change him to the flower that bears his name.
He sings how rushing from the woodland came,
Foaming with rage and pain, a wounded boar,
And young Adonis, loved of Venus, tore.
Herself immortal, yet her power is vain
To shield him, or his ebbing life retain ;
Her prayers and sighs to heav'n ascend, but Jove
Vouchsafes no succour to her dying love.

THE DELUSION OF ATHAMAS.

Book IX.

At the instigation of Venus, Neptune deceives the pilot of Telemachus with a vision, in order to prevent the latter from reaching Ithaca.

AT Neptune's summons from the deep arose
A spirit with deluding power endued;
Deceptive as a dream; yet dreams alone
In sleep beguile, while o'er the waking sense
This spirit of illusion hath control.
He comes, environ'd with a wingèd host
Of falsehoods, fleeting visions, and deceits,
His ever-present ministrants, and o'er
The watchful eyes of Athamas he pours
A magic essence, as the starry vault
He scans, and marks the aspects of the moon,
And bends his looks upon the steepy coast
Of Ithaca, whose threat'ning rocks draw near.
Yet but an empty vision met his view:
The sky above, the sea and shore around,
Were falsely pictured to his sight. The stars

Seem'd to retrace upon the travell'd sky
Their course: Olympus moved by laws unknown;
Nature herself was changed, and while the ship
Forsook her destined port, the pilot still
Beheld it, painted on the empty air.
Still steering towards the shadowy coast, it flew
Before him, till amazement fill'd his mind.
The distant murmur of a busy port
He seem'd to hear, and bade the crew prepare
To disembark upon an isle which lies
Beneath the parent land, that, so deceived,
The traitor crew who, leagued against their prince,
Beset the port, might deem him absent still.
He saw and shunn'd the rocks which crowd the deep
Along that island strand: the horrid sound
Of bellowing breakers fill'd his ears, as burst
The mighty waves around. Anon the shore
Receded, and its mountains, seen afar,
Faded to hazy clouds, like those which fringe
The horizon, when the gleam of sunset dies.
While Athamas the fleeting scene beheld
With wonder, and believed it but a dream,
The east wind blew, by Neptune summon'd forth,
To drive the bark upon Hesperia's coast.
Fiercely it swept the seas, and soon uprose
The destined shore above the western wave.
    Aurora brings the dawn; the stars which shun

Sol's bright approach, as envious of his rays,
Quench their dull fires in ocean, when elate
With joy, the pilot cries: "I doubt no more!
Telemachus, rejoice! In one short hour
Thou may'st embrace Penelope, and find
Thy sire reseated on his island throne.
These joyful sounds from slumber's bonds released
Telemachus; he sought with eager haste
The helm, embraced the pilot, and with eyes
Whose cumber'd lids own'd still the power of sleep,
Survey'd the neighbouring coast. He groaned, alas!
He look'd not on his native land: "Oh! where,"
He cried, "Beloved Ithaca, art thou?
I see thee not; Oh! Athamas, thine eyes
Deceive thee; distant from thy Tyrian strand,
These seas and harbours are to thee unknown."
"Scarce more familiar," Athamas replied,
"Is Tyre to me than Ithaca. My ship
Rides oft upon your waters. Not a rock
Nor shoal but in my memory hath its place.
Yon mountain mole I recognize: behold
That tower-like crag, familiar to thine eye;
List to the roar as boil the waves beneath
Those rocks which menace ocean with their fall.
Minerva's fane thou may'st descry, whose dome
Invades the clouds; and, dimly seen, afar,
The fortress, and the palace of thy sire."

F

While thus he spoke the vision fled ; he saw
Another scene, in truthful hues array'd,
And thus in wonder spoke : " Some hostile God
My senses with a vision hath deceived ;
Before mine eyes lay Ithaca, and wore
Its own familiar aspect ; like a dream,
It vanishes from my bewilder'd sight.
Now, in full view, behold Salentum's towers
And the long outline of Hesperia's coast."

ANTIOPE.

Book XXII.

Antiope was the daughter of King Idomeneus, and the Consort-elect of Telemachus, who speaks the following Panegyric.

ITH grace divine, beneath the linden shades,
She leads the dances of her Cretan maids,
While flutes soft warbling through the groves resound,
And melody and mirth prevail around.
Not lovely Venus more enchanting grace
Can boast, nor brighter charms of form or face,
When forth she moves with her attendant train
Of Nymphs and Graces on the Paphian plain.
   And when with pride and pomp Salentum's court,
In his dark woods to hunt the boar resort,
As Dian's dart unerring is her spear,
Nor Dian wears a more majestic air.
Yet all unconscious of her heavenly charms,
She reins her steed, and wields her sylvan arms,

Nor knows her smile, more dangerous than her dart,
Inflicts a cureless wound, a ceaseless smart.

   When meekly to the temples of the Gods
She bears her offerings, of those blest abodes
Fit habitant she seems; a deity
To whom adoring crowds might bend the knee.
She lays her gifts before each sacred shrine,
And humbly supplicates the Powers Divine
That wars may cease; that tumults may be still'd,
And evil auguries be unfulfill'd.

   She plies her golden needle mid a throng
Of virgins, and relieves their toils with song:
The glory of the Gods her chosen theme,
And grace of heaven to man. He well may deem
Who sees her thus, and hears that lofty strain,
That Pallas, heaven forsaking, dwells with man,
To teach the arts which soften and refine,
And fill the human breast with love divine.—

   Whom gentle Hymen shall unite with thee,
Fairest and best of mortals, blest will be
Beyond his fellow men; yet one alloy,
Else earth were heav'n, will mingle with his joy;
The cruel fear will hover round his heart
That thou, too bright for earth, from earth might'st part,
And he remain, to whom the Fates could give
No future joy, for grief alone to live.

THE APPROACH OF OLD AGE.

MAN'S life is transient as the morning flowers,
Whose buds expand to greet the sunny hours,
But wither as the day declines, and lie
Strewn o'er our path ere evening veils the sky.
Wave follows wave along the flood, and we,
Thus imaged may our own existence see.
Time who or what resists? The wise, the strong,
The valiant, in his course are swept along.
To thee, my Son, rejoicing in thy youth,
Thy happy dawn of life, how sad the truth
That beauty and that strength must pass away,
E'en as the flower which blooms but to decay;
And health and pleasure to thy thought will seem
But the remembrance of a happy dream.
Old age, approaching now, with stealthy pace,
Will grave his track in furrows o'er thy face;
Those graceful limbs enfeeble, and will bow
That lofty frame, erect and haughty now;
Will dry the springs of joy within thy heart,
And bid the present pleasure to depart;

Will fill thy darken'd mind with haunting fears,
And give to pain and grief the remnant of thy years.
Think not the time is far; it comes, 'tis here;—
The inevitable future still is near:
The present we can never grasp; 'tis gone,
'Tis lost, and coming moments hurry on.
Then on the future, not the present time,
Fix'd be thy every thought. Let aims sublime
Confine thy steps to virtue's rugged way;
Let love of justice all thy actions sway;
Thy life be pure, that when the end shall come
The realm of endless peace may be thy home.

THE BARBARIAN AMBASSADORS TO IDOMENEUS.

Book X.

AN olive branch and sword, O King, we bear,
The emblem, one of peace, and one of war.
Choose thou! but peace is our desire: to thee,
For peace, we yield the empire of the sea.
To homeless wanderers from the coast of Greece
We yield the fertile shore, for love of peace,
Blest by the beams of Sol with large increase.
For peace to ease and plenty we prefer;
And life in deserts, or on hills, with her;
Where flowers, the pride of spring, refuse to blow,
And with autumnal lustre, fruits to glow;
Regions of ice, and tracts of endless snow.

*   *   *   *

CLEANTHES AND PHOLÖE.

Pholöe was daughter of the river Liris, and the affianced bride of Cleanthes, who was killed in battle by Telemachus.

Book XX.

HUS slaughter and confusion reign around;
The Daunian troops recede; the ensanguined
ground
Is burden'd with the slain, and there, his eyes,
Darkening with shades of death, Cleanthes lies.
  For him, her guardian and affianced spouse,
Young Pholöe wearies heav'n with prayers and vows:
Fruitless, alas! for on the Daunian Strand
He fought, and perish'd by a Grecian hand.
  With loud laments she fills the Lyrian groves,
And o'er the rugged hills, distracted, roves:
She rends her graceful locks, neglects the flowers
She loved to wreathe, and shuns her fragrant bowers.
The rising morn upon its roseate wing
No solace to her stricken heart could bring;
Nor the pale eve, nor night, the mourner's friend,
A brief oblivion of her cares could lend.

Still, night and day, her tears incessant flow,
And each succeeding hour brought added woe.
The pitying Gods at length vouchsafed their aid,
And to a fountain changed the weeping maid;
A bitter stream springs forth where Pholöe stood,
Strays through the meads, and joins the parent flood;
But still, where'er its winding waters pour,
The sombre cypress shades the barren shore.

THE DEAD BODY OF HIPPIAS.

Book XVII.

EXTENDED on a bier, which silver, gold,
And purple deck, the cold remains behold
Of him, in beauty and in youth by death
Struck down,—whose beauty has survived his breath.
Though closed his sunken eyes, the tender grace
Of youth yet lingers in the pallid face.
Adown his drooping neck, more white than snow,
In sable folds his silken ringlets flow:
Locks beautiful as those which grace the head
Of Atys, or of Jove-loved Ganymed.

  * * * * *

MENTOR ASSUMING THE ATTRIBUTES OF MINERVA.

Book XXIV.

AS darkness flies when o'er the orient mountains
    Aurora leads the chariot of the day,
    While hills and valleys, lakes and sparkling fountains,
  Meadows and forests glitter in her ray ;
So fade the furrows from his aged brow,
    And thus the cloud forsakes his kindling eyes ;
That, majesty and grace illumine now,
    And these reflect the azure of the noonday skies.

And as a tender flow'ret which discloses
    Its opening beauties to the beams of day,
So blooms the Goddess ; lilies vie with roses
    To deck her features with their bright array.
Her martial mien and majesty divine
    Of half their terrors winning grace disarms ;
Softness and sweetness in her aspect shine ;
    Eternal youth unites with never-fading charms.

Her vestures, with a rainbow's radiance gleaming,
  The splendours of the rising morn outvie;
Fairer their hues than sunset glory streaming
  To the dark zenith from the western sky.
Her flowing locks ambrosial odours shed;
  On earth she rests not, but the empurpled air
Upholds her, and sustains her wingèd tread,
  As tempests from his cliff the soaring eagle bear.

She brandishes a lance which warlike nations
  Behold with dread, embattled armies fear;
Her voice, which shakes Athena's deep foundations,
  As sweetest music yet salutes the ear.
The bird of night, upon her lofty crest,
  Sits brooding, shrouded by the snowy plume;
The immortal ægis blazes on her breast,
  And sheds terrific splendours through the forest gloom.

THE SCULPTOR AND THE STATUE OF JUPITER.

From the French of La Fontaine.

A BLOCK of marble which a sculptor bought
So fair, so faultless, such a prize was thought,
That, " Ne'er to common purposes," cried he,
" Shall such a stone as this devoted be.
A God, a God, I'll form, whose hand shall bear
Thunder and lightning: worship, mortals, fear!
Behold, the King of all the Gods is here!"

   The terrors of a Deity invest
The sculptured stone; so truly were exprest
The attributes of Jove, that speech alone
Seem'd wanting to complete the God of stone.
But, ah! he was the first to bow the knee
Before his image, and with dread to see
The gloomy frown, and the uplifted hand,
The forky lightnings, and the threat'ning brand.

Pygmalion thus, by am'rous phrensy moved,
The Venus that himself had sculptured, loved.
And thus the poets of the olden time
Their own creations worshipp'd, and in crime
The sculptor equall'd; thus deluded men
Form creeds and idols of their own, as vain,
And worship the chimæras of their brain.

DISTRUST.

From an Anonymous French Author.

OH that in earthly bower
    There grew, though lowly and of pensive hue,
    Nor in sweet scent excelling, one fair flower,
  Which yet I might endue
With bloom which time and winter might defy;
Might open to the blighting east, nor die,
And long as I should gaze, enchant my loving eye!

  Oh that one summer's day
I might foreknow would fade from painted skies
Glorious as those which caught its herald ray.
  Then with no sad surmise
That storm-clouds might the beaming heav'ns invest,
What bliss were mine, and how serene my breast,
Sharing, in breezy shelter, nature's noontide rest!

Oh would an angel voice
Whisper the welcome tidings in mine ear:
The friend whose truth and zeal thy heart rejoice,
    Is ever thine, sincere,
Unchangeable. Oh then my heart would know
All happiness that friendship can bestow,
And trustful evermore, with generous warmth o'erflow.

    And Love, on which I found
My brightest hopes, had double power to bless,
Did not distrust in fetters hold me bound;
    And mar my happiness.
This joy, methinks, may false or fleeting prove;
And all these dear delights of mutual love
Death may untimely end, or lapse of years remove.

    Those moments since my birth,
On which my friendly star hath shed its beams,
(Though few and distant, as on this sad earth
    Must ever prove the gleams
Of joy which flash upon our night of grief,)
My fears, alas! have render'd dim and brief,
And clouds of doubt obscured those glimmerings of relief.

DISTRUST.

    There are, and happy they,
Wanderers through life o'er dreary paths and lone,
Mid storm and gloom, who welcome on their way,
    With thorns and brambles strown,
Bright buds of pleasure, which anon expand
To bloom, and gladden all the desert land;
Foster'd with faith and love, by hope's sweet breezes
    fann'd.

CAIN.

"And Cain said unto the Lord, My punishment is greater than I can bear."—*Genesis* iv. 13.

I SINK, o'erwhelm'd, beneath Thy dread decree!
An outcast wanderer through the earth to be;
All I have ever loved to leave; to find
My meet companions in the savage kind;
Cut off from all my race, despair my guide;
Remorse and terror ever at my side;
Heaven's intercourse with man forbid to share;
Oh, this is anguish more than I can bear!

I, who Thine image wear, among mankind
The mightiest, yet to ruin am consign'd.
Woe, woe, eternal woe! and who shall stay
The avenger's hand, which shall be raised to slay
The earth-polluting wretch, whose brother's blood
Cried from his reeking altar to his God!
The wilderness unfolds before mine eyes
Its dark, untravell'd paths; waste mountains rise,

And dim, drear forests stretch their endless shade.
Those awful solitudes I seek, afraid
Lest judgment find me here : to them I fly
For shelter from Thy presence; yet the eye
Which saw my crime, will evermore look down
From yon black heav'n, in which I see Thy frown.

THE TRANSLATION OF ENOCH.

"And Enoch walked with God: and he was not; for God took him."—*Genesis* v. 24.

MOURN, Earth, for thou art less like Heav'n;
    thy best,
Thy holiest offspring now is thine no more!
He that hath walk'd with God, with whom the skies
Commun'd, is not; nor doth thy bosom hold
His ashes; he was all too pure for thee.
His soul so sanctified his mortal frame
That it could death withstand, and soar above
Glorious, immortal.
              As he talk'd with God,
Lo! the bright effluence of the Deity
Robed him with light eternal, and a throne
In Paradise received the Friend of God.
    Then triumph, Earth, that thou to Heav'n could'st give,
Stainless, though in thy fleshly vesture clad,
One who now sits among the saints, and holds
Divine communion in the courts above!

THE DEATH OF MOSES.

"And there arose not a prophet since in Israel like unto Moses, whom the Lord knew face to face."—*Deuteronomy* xxxiv. 10.

THE Promised Land! ere close thine eyes
   On things of earth, look up and see
   Thine Israel's home, that Paradise
Foreshadowing which unfolds for thee!
For thou shalt but behold afar,
   From Pisgah's height, yon earthly heav'n;
To thee from perils, toils, and war,
   A surer resting-place is given.

What signs, what wonders since thy birth
   Has Heav'n in human sight display'd!
Portents and plagues affrighted earth
   Has seen, by thee evoked and stay'd.
The Red Sea, smitten by thy rod,
   Its billows heap'd on either hand;
Thou, face to face, hast talk'd with God;
   Thy voice proclaim'd His dread command!

Yet must thou die, and Canaan's soil
  Shall never by thy feet be trod—
Here cease thy glories, ends thy toil;
  Hence shalt thou rise to be with God!
Thy sepulchre, in after time,
  Shall be unknown, but every age
Shall celebrate thy name sublime;
  Thou Leader, Prophet, Warrior, Sage!

And after thee no seer shall rise
  In Israel with thine awful power
Endued—whose prayer could move the skies
  To mercy, or their vengeance shower.
Still Egypt quakes; in memory
  She sees thee wave thine outstretch'd hand;
And Israel weeps, bereft of thee,
  Though gazing on the Promised Land!

C.

THE SONG OF DEBORAH AND BARAK.

"Why is his chariot so long in coming? Why tarry the wheels of his chariots?"—*Judges* v. 28.

WHY do his chariot wheels delay
    To bear him from the war
    In triumph? Lo, his homeward way
  Is lit by victory's star!
The sinking sun forsakes the skies,
    And still he gladdens not mine eyes;
Yet wherefore should I fear
    For Sisera, and that conquering host
Which march'd at dawn for Israel's coast!

And still she gazed till darkness drew
    Its pall across the sky;
When thus, as doubt to anguish grew,
    Her maidens made reply;
He comes not, for with Israel's spoil
    Oppress'd, and faint with warfare's toil,
As Barak's squadrons fly,
    He smites their ranks, nor spares to slay,
Though slaughter'd thousands choke his way.

With Hebrew damsels in his train,
 And warriors captive led,
Eternal glory he shall gain ;
 And thou, apparellèd
In robes whose splendour shall outvie
The Persian woof, and Tyrian dye,
 Won on the battle plain,
In glorious pomp shalt move, and shine
With every treasure of the mine !

Thus at the lattice through the hours
 Of darkness they abode ;
Yet came not those victorious powers ;
 No glittering banners flow'd
In morning's breeze : her heart grew cold
With grief, and thus her fears she told :
 It may be, Israel's God
Is mighty, as His prophet's word
Declares, and hath unsheath'd His sword !

Their records tell of kings cast down
 Who dared provoke His ire ;
Of nations, wither'd at His frown,
 And cities burnt with fire
Which fell from heaven : their God, in fight,
They liken to the sun, in might
 High heav'n ascending.   His attire—

The lightning; tempests from His path
Sweep the awakeners of His wrath.

A vision I behold! our host
   Like scatter'd sheep is flying;
Their chariots broke, their trophies lost,
   Their chiefs in death are lying!
From Harosheth to Kishon's flood
The land is red with Canaan's blood;
   And Israel's tribes, relying
On God, their God of battles, come
To make our land their Promised Home!

THE DENOUNCEMENT OF JUDGMENT UPON DAVID.

"I offer thee three things—choose one of them, that I may do it unto thee."—2 *Samuel* xxiv. 12.

JUDGMENT awaits thee and thy doom
    Thou shalt thyself declare ;
Shall plague or famine fill the tomb,
    Or long, disastrous war ?
Thou, in thy self-relying pride,
Through all thy nations far and wide,
    Hast number'd all who wear
Thy yoke, and made thine impious boast
The myriads of thine armèd host.

Thy God thou didst forget, Whose power
    Avail'd thee more, alone,
Unaided else, in peril's hour,
    Than millions round Thy throne.
He smote the ravening beasts of prey ;

DENOUNCEMENT OF JUDGMENT.

Scatter'd the Philistines' array,
   And led thee, victor, on
Through fields of fight: his shield and spear
Thy safeguard, and thy weapon there!

Put now thy trust in human strength,
   For Heav'n its aid denies;
Thy deeds of sin have roused at length
   The vengeance of the skies!
The angels of the Lord await
His word with plagues to devastate
   Thy land; thy people's cries
Shall rend thy heart, and thou, their king,
Their shepherd, dost the judgment bring!

Thy mighty men, in multitude,
   Like sands beside the sea;
Thy tribes, like leaves which clothe the wood,
   Where will their numbers be
When earth, no more by kindly dews
Refresh'd, her tribute shall refuse;
   And herb and fruitful tree
The desert wind in dust shall lay,
Withering beneath the untemper'd ray!

Or if, by famine's ills dismay'd,
   The swifter plague thou call,

Where then will rest thy hope of aid
    While armies round thee fall?
Then wilt thou trust that helm or shield
Or plates of mail will safety yield,
    Or city's rock-built wall?
The angel of the pestilence
Will set at nought thy vain defence!

Or if the Lord deliver thee
    Into the hands of men,
Thy towers will sink, thy champions flee,
    And thou in cave or den
Wilt hide the head which Israel's crown
Hath worn; thy splendour and renown
    Lost, as they had not been!
Thy people captive, and their land
    Laid waste beneath the spoiler's hand.

THE XCVII. PSALM.

GOD'S POWER AND JUDGMENT DISPLAYED IN A THUNDER-STORM.

EHOVAH reigns—His sacred Law
Let earth receive with joy and awe,
And to her utmost isles be glad!
The Lord, in clouds and darkness clad,
With righteousness surrounds His throne,
And heav'n and earth His judgments own.
Before Him in avenging fire
They perish who arouse His ire.
His lightnings shed their awful light:
Earth sees, and trembles at the sight.
The mountains smoke, the rocks descend,
The mighty hills their summits bend,
And melt like wax before the flame.
The heav'ns His righteousness proclaim:
His glories all the world illume,
And nations gaze and dread their doom.

Confounded now, ye sinners, be,
And who to idols bend the knee;
Bow down, ye Powers, before His throne,
And worship him Who reigns alone!

 Zion, exultant, hears the sound,
Thy voice in thunders roll'd around.
The daughters of Judæa sing
Thy judgments. Thou, creation's King,
O'er earthly gods exalted high,
Supreme, almighty, now art nigh.

 Oh ye, who in this awful hour
Through love, can triumph in His power,
Hate sin, for God preserves the pure;
His people are alone secure.
But thou, the tyrant, thou, the unjust,
Be humbled your proud heads in dust.
For you these fiery portents shine;
For you awakes the wrath divine!

 Rejoice, ye righteous, for the Lord
Prepares in heav'n your bright reward;
Amid the gloom shall glory rise,
And light celestial glad your eyes:
Your thankful hearts shall ever bless
His mercy, truth, and holiness.

SOLOMON'S PRAYER

AT THE DEDICATION OF THE TEMPLE.

"But will God indeed dwell on the earth?"—1 *Kings* viii.

THOU Who hast said, Around My throne
Thick clouds and darkness I have thrown;
For Whom the heav'n of heav'ns in vain
Unfolds, Thy presence to contain;
Wilt Thou for man the skies forsake,
And earth for him Thy dwelling make;
This house, which I have built for God,
Shall it, indeed, be His abode?

Yet from Thine everlasting seat
Behold us, kneeling at Thy feet;
And may Thy gracious ear approve
Thy suppliant's prayer, that from above
Thine eyes may shed their sacred light
On this Thy temple day and night;
For Thou hast promised, by Thy name,
To give it glory, riches, fame.

This house we consecrate to prayer:
Oh, be Thy mercy ever here!
And when within this holy place
Thy people kneel, forgiveness, grace,
And succour grant. Should foes assail,
And Israel's hosts in battle fail,
Then from this altar as our cries
Ascend, do Thou in might arise!

Should famine stretch her mildew'd hand
O'er the bright valleys of our land;
Should drought, or heats, or with'ring frost
Lay bare our plains; or on our coast
The locust from the sunless sky
Descend, or pestilence be nigh,—
Let Israel then before Thee bow
In these blest courts. Have mercy, Thou!

Or if in distant lands we mourn,
And languish, hopeless of return,
Yet when we turn a prayerful face
To Thee, and this most holy place,
And humbly kiss Thy chastening rod,
And bring our hearts again to God;
Hear Thou in heav'n, and set us free,
Again, in peace to worship Thee.

A REVERIE IN CUMBERLAND.

Mind-Pictures Realized.

WHO that beholds, at twilight's mystic hour,
    The star-lit beauties of a pastoral vale,
But sees, mind-pictured oft, in stream and bower
    The woods and waters of some classic tale?
  There slumbers Tempè; the Elysian plain,
Or Happy Valley, with its shades serene:
  Yon woodland is the Faëry Queen's domain,
That glittering lake the dwelling of Undine.

But in a mountain desert Fancy plumes
    Her daring pinions for adventurous flight:
Each outlined steep familiar form assumes,
    Each rock, remember'd hues, as to her sight
In day-dreams oft they rise; the imaged scene
    Where warriors wild, dread phantoms, hermit saints,
Floods, earthquakes, or volcanic fires have been,
    As legends chronicle or fiction paints.

Love, too, has flourish'd in a mountain clime,
    As many a tender tale of joy or woe,
Of later days, or of the olden time,
    Which rocks and hills to memory bring, may show.
His voice still whispers on the mountain air ;
    His form the shepherd wears : the cedar grove
Has bent its boughs o'er many an amorous pair ;
    Has sigh'd with those who mourn'd their hopeless love.

In yon pine-crested peak, which soars afar
    Into the twilight blue, behold the bower
To which pale Luna guides her crescent car ;
    Where blest Endymion waits the ev'ning hour.
And mortal loves such scene may consecrate :
    Ye hapless pair who in the Afric isle
Lived, loved, and died, to mourn your world-wept fate,
    I wander oft where mountain gardens smile. .

What pensive thoughts are his, whose lonely way
    Winds through a wilderness of barren hills !
At each new aspect of their dark array
    His soul a wilder inspiration fills.
He dreams, sole tenant of the youthful earth
    Himself, the rugged scenes around him spread,
Gloomy and vast, are Nature's second birth ;
    The deluge ocean's late forsaken bed.

Or should the Storm-Fiend work his vengeful will,
  On howling wings the desert defiles sweep;
Hang his black banners round some kingly hill,
  And mutter to the waste in thunders deep,—
He thinks on Sinai : as the lightnings glare,
  And rolling echoes strike the wild with awe,
He sees the hosts of Israel marshall'd near,
  He hears the Voice Divine pronounce the law.

And softer scenes alike their memories bring
  To cheer him, toilworn, on his rock-strewn way;
See sunset o'er yon hills its radiance fling,
  Whose summits with the hues of heav'n are gay.
On Christian's gaze, Despair's dark boundary past,
  Thus burst the mountains of Immanuel's Land;
On whose bright peaks the gate of glory cast
  Its rays; their slopes by Beulah's breezes fann'd.

A YOUNG LADY'S FAREWELL TO PETERBOROUGH.

ADIEU, belovèd town; with mournful gaze
I view thee, fading in the horizon's haze!
Farewell, dear City! tears bedim mine eye,
As borne from thee,[1] too swift, alas! I fly.—
How many happy, ah! too fleeting hours
Have fled since first I view'd thy welcome towers.
Three moons have waned since that delightful time
When broke upon mine ear St. John's sweet chime.
Then hope and pleasure revell'd in my breast;
But now, with melancholy thoughts opprest
Of pleasures past, and valued then too low,
How shall I check the current of my woe!

Let me, while yet the privilege remains,
Fix my last look upon these smiling plains;

---

[1] In the up express.

Throw my last glance on yon majestic fane,
And view thy windings, "willowy Nene," again.
Too transient prospect! rushing swiftly by,
The well-known objects vanish from my eye;
The fields, the woods, the city, and the stream
Melt like the fleeting scenery of a dream;
Yet deeply imaged in my heart shall dwell
The Elysian scenes to which I bid farewell.

No more my feet along Thorpe-walk shall stray,
Nor seek the Stanground, nor the Woodstone way.
Fletton, farewell! no more thy tower I view,
And railway stations, you I bid adieu;
Infirmary—I speak the name with awe—
Abode of horrors, lancets, scalpels, saw,[1]
Crescent and prison, Union-house, Thorpe Hall,
Bridge, Gravel-walk, adieu, adieu to all.
Yet in my grief one solace can I find;
The thought is balm to my desponding mind
That duty calls me hence. I go to rule,
As Lady Paramount, our village school.

---

[1] Miss L——'s original words were—

"Infirmary—at that dread word I pause—
Abode of horrors—doctors, lancets, saws!"

To free benighted children from the fetters
Of ignorance, and make them learn their letters.
To this dear duty I devote my hours,
How tedious else, and consecrate my powers!

CHISWICK PRESS:—C. WHITTINGHAM AND CO., TOOKS COURT,
CHANCERY LANE.

www.ingramcontent.com/pod-product-compliance
Lightning Source LLC
Chambersburg PA
CBHW020336090426
42735CB00009B/1552